Ali Ben Aouiene

Energy Efficiency, Which Way to Go

Ali Ben Aouiene

Energy Efficiency, Which Way to Go

Energy Performance

ScienciaScripts

Imprint

Any brand names and product names mentioned in this book are subject to trademark, brand or patent protection and are trademarks or registered trademarks of their respective holders. The use of brand names, product names, common names, trade names, product descriptions etc. even without a particular marking in this work is in no way to be construed to mean that such names may be regarded as unrestricted in respect of trademark and brand protection legislation and could thus be used by anyone.

Cover image: www.ingimage.com

This book is a translation from the original published under ISBN 978-613-9-51297-3.

Publisher:
Sciencia Scripts
is a trademark of
Dodo Books Indian Ocean Ltd. and OmniScriptum S.R.L publishing group

120 High Road, East Finchley, London, N2 9ED, United Kingdom
Str. Armeneasca 28/1, office 1, Chisinau MD-2012, Republic of Moldova, Europe
Printed at: see last page
ISBN: 978-620-4-44744-5

Table of contents:

Energy Efficiency The Way Forward

The Expert: Ali BEN AOUIENE

Preamble:

Energy Efficiency is a phrase that everyone talks about, but does everyone know the paths to achieve this efficiency? And are they all equally efficient in their paths?

In this book and as an expert in the field, I will try to shed light on the best ways to preserve energy and consequently preserve our planet.

To be successful in energy efficiency work, you need to master all the tools to be used.

This theme is on everyone's mind, both individuals and country leaders who constantly talk about it, but are they following the right paths to achieve their goals?

I. Introduction

In this publication, I have tried first of all to define the main areas of energy saving that individuals or companies should seriously consider.

Then, I will focus the light on the key figures that we do not usually talk about, then, I will approach the subject of units and energy parameters, (especially those that represent nuances and confusions) and finally I will give a list of actions to follow for a rational use of energy.

"Of these actions, I can predict that a few will be heard for the first time."

The details of all the energy saving actions listed will be the subject of other books for those who can and want to deepen their knowledge.

II. Pathways to Energy Performance

Energy efficiency is in other words, the performance in the consumption of energy of an equipment, a process, an establishment or even a country and for what not the globe.

The mission and work that targets energy performance must follow at least one of the following axes and at best, follow all of them:

a) **Optimization and improvement of the energy consumption** of an equipment or a system that necessarily consumes at least one of the multiple forms of energy namely: Electrical, Kinetic, Mechanical, Thermal, Potential...etc.

b) **Recovery of energy** and which consists in recovering a part of the energies lost in the processes and the energy systems, these losses represent rates, unimaginable, for a great number of us.

c) **Energy reuse**, in fact we are not familiar with this term, but we hear more and more about the Circular Economy which is part of this principle, but in my opinion, the latter is only a part of what I call energy reuse. And you will see in what follows, that man actually throws a large part of the energy into nature, apart from the known losses (avoidable and/or unavoidable).

Each of these axes can also be subdivided into three paths and this is what I will detail in what follows:

I start of course by detailing the above mentioned axes.

1) Optimization and improvement of energy consumption.

We begin with this first axis that, almost, in the whole world there are organisms and regulations that frame the works of rational use and control of the energy consumption.

I am not going to detail the methods followed to rationalize the energy consumption, but I will try to give the ways to follow to ensure it.

a) The first way is to think, from the beginning of the design, purchase and installation of any equipment or energy system, to the performance of the same.

Concerning the choice of the equipment which are part of the energy systems, I can say that there are already orientations towards that, (Example the reference IS050001, SMEn) but given that in general, the price of the equipment increases with its performance, this way remains less travelled, the overcosts can reach more than 20 % and 30 % of the cost of a standard equipment what makes that the decision-makers pass, the majority of the time, beside this way

However, another important piece of information that is ignored by some of us and in particular the decision makers, or may be neglected by others, is the cost of the initial investment, which is sometimes negligible compared to the overall cost spent on the equipment over its lifetime.

(We don't know how long it will last and sometimes we don't talk about it).

In practice, (In general), there are three expenditures on equipment:

- Initial cost (investment at purchase)

- Operating cost (which includes maintenance, replacement of spare parts, etc.).

- The cost of energy consumed by the equipment over its lifetime.

I give in the following, some figures that some will be surprised, and as an example:

* For electric motors, which already account for about 80% of the power of installed electrical equipment in industry, energy consumption expenses over their lifetimes represent about **95%** of overall expenses.

* For air compressors, energy costs represent about **75%** of the total costs, the initial investment represents 12 to 13% and the rest is reserved for maintenance.

(The consumption of compressed air in the industry varies from 5% to more than 35%, with an average of 10% to 15%).

* For the lamps, either incandescent or fluorescent, the expenses on energy reaches 90% of the global expenses on these lamps.

A small balance sheet and a little quick calculation shows us that the additional cost of purchasing efficient equipment represents little, compared to the expected savings on energy consumption.

I give the possible rate of energy performance improvement for these three examples:

Example 1: Case of the electric motor

The extra cost for an IE3 efficiency motor against a Standard motor is about 20% of the purchase cost which represents less than 5% of the expenses on this motor over its lifetime, so it will be 1% of total extra cost, while the energy efficiency will be improved only by 1%, which will be a gain of 1%*90%=9% of the global cost (Balance gain of 8% on the global expenses over the lifetime)

Example 2: Air Compressor Case

The extra cost for a compressor equipped with an EVG (Electronic Variable Speed Drive) against a standard compressor (On/Off) is 30% to 35% of the purchase cost which already represents about 13% of the global expenses over its lifetime, so it will be 5% extra cost on these expenses, while the energy efficiency could improve 20% and more, which will make a gain of 20%*75%=20% of the global cost (Balance gain of 15% on the global expenses over the lifetime of the compressor)

Example 3: Case of incandescent lamps.

The extra cost for a LED lamp against an incandescent lamp is about 1000% of the purchase cost, which is less than 5% of the total expenses over its lifetime, so it will be 50% of total extra cost, while the energy efficiency can be improved by 550%, in addition to the lifetime of LED lamps that reaches 50 times more than that of incandescent lamps, which will make a gain of 550%*90%=495% of the overall cost (Balance gain of 445% on the overall expenses over the lifetime without counting the lifetime).

Therefore, there is reason to think seriously about the design, the life span, the technology, the purchase of more efficient equipment, the gain is always justified (Expert's Opinion).

Another branch of this first path of energy performance, although it is the twin sister of the 1st branch (good choice of the system), is the technical study of the installations that must take into consideration the energy efficiency too.

In the same way, the extra cost of an energy efficient installation is justified in front of the energy gain and sometimes, it allows to solve even some technical problems, examples are presented below to clarify the ideas.

Let's take the example of the electrical installation, the electricians speak about regulatory or normative voltage drop, which for example for an industrial installation and concerning the power part (motors and others), the allowed voltage drop (Maximum) is of 5%, and what it should not be forgotten it is that the drop is synonymous of power loss, that is to say 5% of energy loss only in the electrical cables. (5% of the electric bill too).

So, if we aim at a voltage drop of 3%, it will be better than 5%, we will gain 2% on the electric bill (***the extra cost is paid once, the gains will be forever***).

Similarly for fluids and in particular compressed air, although design offices are not familiar with the recommended pressure drop of 5%, which is already about 3.5% of power loss for the compressed air station, In practice, a pressure drop of between 5% and 10% of the operating pressure is aimed at, but we note that a 5% loss is much better than 10% (the gain on this item will be a minimum of 3.5% over the life of the plant).

Also, for the temperature drops on the surface of the pipes carrying hot fluids, in practice, it is necessary to ensure a good insulation to reduce the gap between the hot surface and the ambient air (Attention here, there are dimensions of economic lagging, that is to say, even if we have lagging in stock in our stores, we must not exaggerate, otherwise we pass on the other side of the zone of energy loss).

Therefore, in general, it is necessary to minimize the pressure drop, the voltage drop and also the temperature difference between the surface of the pipes and the environment by increasing the cross-sections of the pipes, cables and heat insulators.

b)   The second path is the proper use of existing energy equipment and systems.

In fact, a free and very economical action must become part of everyone's habits, namely: turning off unused equipment and I will give in the following several practical examples:

Light, in workshops and garment factories for example, and during the lunch break, which lasts one hour and already represents 12.5% of the factory's operating time for an 8-hour shift, during this time it is strongly recommended to turn off the lights, the gains will be 12.5% of the Lighting item, (Almost without investment).

It is true that this point is obvious to all, but I assure you that it is not part of our habits, so another point to consider is that it is ignored by many operators, and that of the auxiliaries of the machines that remain in function even if the machine is stopped. (For example, machine cooling pumps).

Another point that can be found in many industries is the centralization of utilities, I always give the example of pumping water or thermal oil or other fluid. In general the dimensioning is done for the totality of the served machines, on the other hand these last ones are not always in function simultaneously, whereas the utilities are put in function at 100% and as long as the plant is in exploitation, the conceivable solution is the fine regulation and the servo-control of the used capacity of these utilities according to the number of machine in function and also according to the capacity of these last ones which is in general variable

And many other points that can be considered.

c) The third path is the energy audit and expertise mission, which is already regulated in many countries.

Incentives for performing audit assignments and grants and incentives are provided for performing these assignments.

In general, this mission consists of a procedure that begins with preliminary visits for a diagnosis of the energy situation, realization of balances and calculation of the efficiency of equipment and energy systems.

Then, by comparing these efficiencies to reference values or usual values, conclusions and recommendations will be drawn for further implementation.

This mission requires knowledge in the field and to succeed in this mission, the auditor must have in-depth knowledge (especially on the practical side) in all technical areas. (Study, Maintenance, Operation, Design, Functioning of the equipment, Accessories, Regulation...etc.), as well as a knowledge of all the energy parameters (Temperature, Pressure, Humidity, Speed, calculation formulas,...etc.) and most importantly the interaction of all these parameters with the specific case of each organization and each equipment.

I mention this because some energy experts and even listeners are already confused about the units of these parameters, and you will be convinced of this in the following, when talking about these units.

But before that, it is necessary to be able to estimate from the beginning, (as a preliminary) the potential of energy saving and to justify it afterwards. In fact, some of us think that by switching on any equipment, it will consume what it needs.

It's true and it's false, the true is that the equipment will consume the right amount of <u>final power</u>, I give an example of a roller mill that is equipped with an electric motor, the power needed in fact, it is only the <u>power useful</u> to grind, ie at the level of the cylinders. However, these are linked to pulleys and through belts, are linked to another pulley at the level of the motor shaft, this motor provides mechanical power and consumes electrical energy.

The fallacy is that this absorbed power, (Electric) at the level of the motor, is not always necessary because there are always losses at the level of each accessory (Belts) and equipment (Motor) and consequently the degradation of the system efficiency.

So, everyone knows the amount of energy consumed (**power input)**, but not sure that everyone knows the **useful** energy of his process. And it is up to the auditing work to detect the losses and the difference between <u>the useful</u> and <u>the absorbed energy</u> and then to look for the solutions to be considered to reduce the absorbed energy and not the useful one.

In this specific example (roller mill), I give you some hints, the first is the tightening and the good tension of the belts, also the number of belts, if the manufacturer envisages several belts, it is necessary to take care not to reduce this number, otherwise there will be lack of transmission between the shaft of the engine and the shaft of the cylinders thus, the engine will provide a mechanical energy which will not be transmitted to 100%, therefore a part will be lost (and it is why one hears sometimes a specific noise with the slipping of the belts)

Another track is the load rate of the motors, in the case of oversizing of the latter, (and this is the general case). Knowing that the losses are inversely proportional to the load rate of the motor, therefore, there will be more losses if the latter is under operated (because many losses are fixed and do not depend or little on the load).

And to get a good grasp of this point, I give examples of the usual yields of some equipment:

- For electric motors,

- if the motor operates at its nominal capacity, which in practice is very rare because when sizing, there is always a safety margin of 20 to 25%, the efficiency varies from 75% for low power to 90% for high power (on average, we are talking about 85 to 87%).

- if the motor is operated at low load rates, the efficiency will be degraded, for example, if the load rate is 25%, the efficiency loss will be more than 15%.

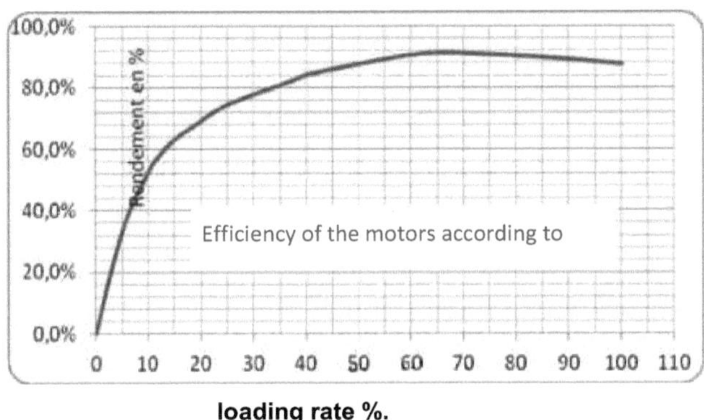

- **For air compressors,**

- The useful power is only 20% of the power absorbed by the engine, all the quantity lost is found in the cooling radiator. (That is to say in the form of heat), in fact the compression of the gas increases the temperature considerably and this according to the law: **"P*V=nR*T".**

- Another characteristic, for the screw compressor, the technique makes that the compressor must work in load at most 75% of the time, during the remainder of the time, the engine works with no load, which makes in addition, almost the 1/8th of loss. (The motor runs at no load when there is no air production).

- **For power plants,**

- The steam power plant is called Steam Turbine (ST) and fortunately the number of STs, leaves more and more room for other types, this type of power plant has a yield of about 38%, which means that at the level of these power plants, 3 / 5 of the energy absorbed are thrown into nature!

- For the gas turbine (GT) power plant, the efficiency climbs but in practice hardly reaches 45% to 50%.

- Whereas for the power plant that already recovers part of the hot gases from the CT, which is the Combined Cycle (CC) and which in fact combines CT and TV, the efficiency is already at most 55%.

It is true that 1% gained is a lot at the level of the plants (billions of euros) but the real losses are much more than that.

- **For steam boilers,**

- We have a first transformation at the level of the combustion of the fuels whose output is already in the best cases, from 90% to 95%.

- Then there are the losses through the walls which reach 2% of the nominal power (nominal operating point).

- Then, the losses by the purges of deconcentration and which depend on the quality of the feed water and the mode of deconcentration, they reach sometimes 10%.

- The flash steam following the return of condensates to the food tank, which reaches a loss of 5% and more. (Especially for atmospheric tanks).

- There are also the losses at the stop, in fact at the time of the stop of the burners, the boiler cools down and we must recover this loss during the operation of the burner, the rate of these losses depends on the regime. (A practical case gave about 10% of losses).

- This gives an overall steam boiler efficiency of 75% or less.

- **For refrigeration units (with refrigerant gas),**

- We speak of a COP (Coefficient of Performance) and not of efficiency, the COP varies from simple to double between summer and winter (in the Mediterranean area), we speak of cold COP from 2 to 2.5 which means, consuming one unit at the electrical level (Compressor), we will have 2.5 units at the level of the evaporator and 3.5 units at the level of the condenser (hot COP).

- **For centrifugal pumps and fans,**

- The efficiency of pumps and centrifugal fans varies according to the brand, and the power range is from 55% to 75% at best, not to mention the efficiency of the motors that drive them.

- This already low efficiency varies negatively each time we move away from the nominal operating point of the equipment.

This is why the potential for energy savings is so important and why we should seriously consider reducing these losses, recovering part of them and why not reusing another part (this is what we will see, in the following, with the energy efficiency axes).

In practice, many establishments have worked on this subject, I give two real examples of success:

- **A factory operating in the Para pharmaceutical field**

This factory **managed to reduce its bill by 40%,** on its own, that is to say without the intervention of an external expert. However, after the intervention of an external expert, it managed to do more, and reduced the new bill by 20% more.

It's no secret that I'm going to give the leads followed by the factory:

- Implementation of an energy accounting system (Electricity, Chilled water flow, Temperature....etc.)

- Focus on the most important consumption item which is the cold (the plant and in totality is air-conditioned), this item consumed more than 45% of the bill.

- Reduction of compressed air leaks, after quantification following the installation of global and workshop compressed air flow meters.

- And since leaks will never be zero, they have also opted to install solenoid valves to close the circuits of the workshops outside of manufacturing.

- Installation of Variable Speed Drives on all pumps and fans, it is worth mentioning here, the potential gain by installing variable speed drives. (For centrifugal pumps and fans, which is the case for the majority of such equipment worldwide, the motor power varies as the cube of the speed, in fact the flow rate varies as the speed and the pressure varies as the square of the speed and therefore the power varies as the cube of the speed (or flow rate). **Pu=f(Pressure, Flow...)**

In other words, if we could divide the used flow by 2, so the necessary speed will be half, the power will be reduced by 87%, something that may not be possible in some cases, but a variation of flow of 10% less, reduces already, the absorbed power of about 30% (Something that is very possible with the variation of the load that is very likely).

- **A second factory: Imprimerie Privé, High level.**

In this printing plant, the auditor was able to extract maximum benefit and **save the facility 35% on its energy bill.**

Also I give in what follows the tracks and actions:

- Reduction of the subscribed power, it is the contractual power with the energy distributor which requires a fee to be paid each month by the establishment.

- Optimization of the power factor, this power factor gives the companies 5% of the energy price if it reaches the unit.

- Replacement of T8 fluorescent tubes (36W) with T5 fluorescent tubes (28W), for an equivalent luminous flux, or even better, the gains on this item are more than 30%.

- Installation of a variable speed drive on the air compressor motor, which reduced the consumption of this item by 52% and the energy bill of the printing plant by 9%. (Gain on the idling of the compressor which was enormous).

- And many other actions.

Conclusion:

I believe that the first axis of energy efficiency is the right choice of designers (who take into account energy efficiency), then the right choice of equipment and installations (even if they seem more expensive) and then the right operation of this equipment once installed and finally by expert assessments and energy audits to detect drifts in energy performance and improve or correct it if necessary.

2) Energy recovery.

The second axis of energy efficiency, although it is already in the mind of manufacturers and operators of energy equipment and facilities, there is still, much work to do.

In fact, the recovery of energy is in the majority of cases, requires too technical interventions, on equipment and facilities that have energy losses.

Sometimes this requires large investments and in other situations it will be necessary to add other equipment and/or other processes.

In the same way, I will detail in the following, some paths to follow to recover lost energies.

First of all, we have to agree on the notions of recovery and reuse, we understand by recovery of energy, the exploitation another time of a part of the lost forms of energy, (Exp: recovery of the heat of the purges of the steam boilers) whereas the reuse consists in exploiting another time, a part of the energy already used by the equipment or the energetic installation (Exp: to reuse the water of the last rinsing of washing machine for the first washing of the following cycle).

a) Recovery of lost energy

The first path to follow and due to the (more or less low) yields of the equipment, as already mentioned above, is to recover the lost energies since they have already cost us a lot of money.

I give, first of all, in what follows, the forms and processes of energy recovery that are already known and used.

* Cement factory

I will start with the very important deposit that is recovered from the most energy-intensive processes in the world industry, namely cement plants. In fact, I will quickly remind you of the phases of the PORTLAND cement manufacturing process, which are

- Extraction and transport of raw materials from quarries to the plant (Limestone, Marl, Clay, Sand, Iron ore).

- Grinding of the flour components after dosing to obtain CRU flour in mills (mostly horizontal roller and ball mills).

- Storage of the flour in silos before passing to baking through recovery cyclones (called precalciner).

- The pre-calcination consists of a first chemical reaction, namely the decarbonation of the material (release of CO_2 gas), this release requires a temperature of more than 400°C which is recovered from the cooking oven

(Energy losses from cooking in the oven).

- The cooking of the decarbonated material and which consists in transforming the material into clinker (main material for the manufacture of cement), this transformation requires energy (It should be remembered that a clinkerization reaction occurs in the kiln and is exothermic, that is to say it releases heat contrary to the decarbonation which is endothermic).

- The product at this stage is clinker which is already a merchant product, but, it is the main product for cement production, this product will be stored after cooling in the cooler, the cooling air after being heated, it will be recovered as combustion air. (Called secondary air and tertiary air).

- The last phase, is the manufacture of cement from clinker and the addition of limestone and clay according to the quality required and the type of cement desired.

To conclude, and fortunately, during the manufacture of cement, a large part of the energy lost during the different phases of transformation is already recovered.

- **Brickworks**

The second important deposit which is also recovered at the level of the energy consuming processes of the industry, namely the brick factories, in fact, I quickly recall the phases of this process of manufacture of hollow brick (Baked), it is generally about five phases:

- Extraction and transport of raw materials from the quarries to the plant (different types of clay, sand).

- Preparation and first grinding of the components, after dosing to have pretreated clay and this in horizontal roller mills, it is possible that there will be other types of machines such as: grinding mill, rougher...

- Bricks are manufactured by a finishing mill, usually a roller mill, a mixer (with water) and a moulder (Extruder) to finally have green bricks cut by multiwire machines and ready to be dried.

- Drying of bricks in two-row swing dryers (or even four-row dryers, but this is rare). The thermal energy consumption of the dryer is almost 1/3 of the consumption of the brick factory.

- The cooking which is done after drying and this in a tunnel kiln of a few tens of meters, 3 parts constituting the kiln, namely: the zone of the pre-firing (preheating), the zone of cooking (Baking) and the zone of cooling. The consumption of the furnace is 2/3 of the consumption of the plant. It must be said here that the cooling heat is recovered and already represents almost 20% to 25% of the energy consumed in the kiln.

The recovered energy is used, usually in the dryer, (injected into the mixing chamber).

- **Steam boiler**

Another known recovery system, mainly by the operators of large steam boilers and in particular those used in power plants, is the recovery of part of the waste heat from the chimney to heat the boiler feed water.

This loss at the chimney level represents a relatively low rate, 5% to 10% in bad cases, but given the power used and the energy consumed, this action is always beneficial and also profitable.

- **Air handling unit**

Another process used to recover part of the lost energy, although it is not very well known and not exploited, is the recovery of the energy extracted with the air rejected at the level of the CTA (Central Air Treatment).

The operation, in general, is done in the following way:

- Fresh air is brought in from outside and mixed with the return air to be conditioned through the hot water and chilled water coils (as required).

- Air taken from the air-conditioned room and whose temperature is equivalent to the set value (desired value in the room) which is different from the temperature of the fresh air (higher in heating mode and lower in cooling mode).

- Blown air after treatment and conditioning to be injected into the room.

- And to balance the air inlets and outlets, a quantity equivalent to that of the fresh air must be discharged and the temperature of the return air.

The heat recuperator, when it is foreseen, consists of an air/air heat exchanger inserted between the fresh air (cooler) and the rejected air (warmer) to recover important temperature differences (a few degrees).

On the other hand, we note that other processes are too energetic and represent huge losses and whose recovery is absent.

I give in the following, some examples which I believe are the most important and energy consuming:

* Power plants

As already said, the yields of the power plants are considered among the lowest, so they have the most important losses and as these plants supply all the end users of the energy, so the monetary losses are difficult to limit.

But unfortunately, these losses are not recovered, it is true that the principle of recovery requires two important technical parameters:

- The temperature potential, because it is not enough to have a lot of energy (kJ) but it is necessary to have sufficient temperatures to use them.

- The usefulness of this recovery in the establishment itself, it is necessary to need this energy, otherwise why recover it and what to do with it?

It is true that the governments of several countries are thinking about this recovery and that is why some countries have exported part of the production of electrical energy to third parties via cogeneration and tri-generation for example, in fact it is a matter of producing electrical energy locally (in factories and tertiary establishments) and recovering part of the thermal losses, of course, these establishments must need heat and / or cold to benefit.

The yields of these small power plants therefore increase to higher rates, up to 60% to 80%.

Let's go back to the power plants, it's true that they don't need the available heat, but on the other hand, they can produce (as the cogeneration units) heat and cold and export them to other establishments such as: hot water, sanitary hot water and chilled water which are very much used, and especially cold since it consumes electrical energy, and therefore the gains on a national and international scale will be very important.

- Refrigeration units

Refrigeration units consume mainly final electrical energy (do not forget primary energy with power plant efficiency of 50%).

The share of cold in the world consumption is estimated between 15% and

25%. In fact, cold is everywhere, in our homes, in our cars, in our stores and stores ....etc.

These refrigeration cycle units (generally) operate according to the following principle:

- Compression of refrigerant gases from low pressure (LP) from the evaporator to high pressure and back to the condenser. Here the work of the compressor and therefore the energy consumed depends on the compression ratio (Ratio=HP/BP), the consumption is all the more important as the ratio is greater.

- Condensation of gases to pass to the liquid state which requires fresh air or fresh water to cool the condenser (the air or water at the exit will be heated). Here the role of the condenser is primordial and is the passage from the gaseous state to the liquid state.

- Expansion of the cooled liquid to be able to evaporate it afterwards, the expansion is done from HP pressure to BP pressure.

- Evaporation of the refrigerant liquid which is already cooled and expanded in a liquid/air or liquid/water exchanger, in order to evaporate a liquid, the air or water must be warmer than the liquid.

In fact, it is the principle of evaporation of fluids which is done according to the PV=nRT law, in other words and as it is known by the readers that water at atmospheric pressure can only evaporate from 100°C, this same principle applies for refrigerants which evaporate at 1 bar for example at negative temperatures (Characteristics of fluids).

And this phenomenon is exploited in evaporators to cool air or water.

Up to now, we have not talked about recovery, but we must know that in refrigerating machines, we only need cold and whose efficiency is quantified by the Coefficient of Performance: COP cold (COP= Efg evaporator / Eelc compressor) and we must also know that the energy available at the condenser is much more important than that recovered at the evaporator.

(E condenser = E evaporator + E compressor)

So, we have more energy at the condenser than we need at the evaporator, **why don't we think enough to recover and use it?**

It is true that the refrigeration machine is reversible and therefore can be used to produce heat or cold. But the habit of the designers (engineers and design offices) has not thought of using both forms of energy at the same time, although nowadays with the VRV System (Variable Refrigerant Volume), it is easily possible to exploit both forms.

* Air compressors :

For the air compressors and as the compressed air station consumes between 5% and more than 35% of the electric bills of the industrial companies, and as the efficiency is the lowest, about 20%, therefore we have 80% of energy lost in the form of heat at the level of the air compressors.

Also why not think about getting it back?

Yes, the temperature is high enough to be used to heat air or water to temperatures of 40°C or 50°C, the exhaust air temperature is between 80°C and 90°C.

This potential and many others, encourage us to recover them well.

It must be said that the intervention on the equipment to recover these energies requires technicality and in the majority of the cases, require the agreement of the manufacturer, not to disturb the functioning of the machines.

However, and already the manufacturers of compressors have integrated recovery devices but never or can seen installed in the industry.

b) Energy recovery from waste

We pass to the second very important way also and that consists in recovering the energies stored in the thrown products.

We know that any combustible product (Possibility of catching fire) stores calorific potentials sometimes very important such as household waste.

But there are many others, such as :

- The pieces of wood,

- Used cardboard,

- Used tires,

- Oils,

- Plastics,

- Li liqueur noir, product of the cooking of Alfa

- Any other product except metals.

The most used process is the incineration of managerial waste, which allows to recover the calorific value of this waste, to produce a quantity of mechanical energy and consequently an electric energy easy to transport.

But much remains to be done especially, the recovery of other types and which can be specific to the various industrial processes as for example the following real cases:

- Incineration or combustion of wood waste in an MDF wood manufacturing plant, the wood dust is burned in a recovery boiler to heat thermal oil.

- Combustion of black liquor, resulting from the cooking of ALFA paper pulp, this liquor has a high calorific value and is concentrated in concentrators and then used as fuel in a recovery boiler.

- Recovery of waste tires, which are shredded and introduced into cement clinker kilns, the calorific potential is too high.

Therefore, we must ensure that all waste is recovered, provide technical possibilities and encourage industries to use them, there is no lack of experience.

c) Increased equipment life

In fact, it is a very important parameter that can be, the majority do not think of it although it is commonly exploited which is the parameter time, the consumed energy it is only a power absorbed during a defined time, to reduce the energy it is equivalent to reduce the absorbed power, or the time of use or both at the same time.

Reducing power is the work of energy efficiency, reducing the time of use is the concern of any operator or at least anyone who has an energy bill to pay.

On the other hand, increasing or extending the life of the equipment by making the product last longer (therefore more robust, more technically efficient), or the proper use of the products by the operator (Good Maintenance, Good Operating Conditions) so that they last longer, will save energy, especially on a national and international scale.

In fact, to last longer is to produce good quality but to produce less in time, thus spending less energy.

The manufacturers may not agree, but we can compensate for the loss of revenue by increasing the selling price and it is worth it for the buyer since the product is more efficient.

And that's what we do, we try to buy good quality even at higher prices.

This requires further study but I believe it is a good path to take.

Conclusion:

The second axis is the recovery of lost energy and as mentioned in this section, these losses are, in many cases, more important than the energy consumed.

This axis requires technicality, money but we must not forget that free energy is available and is not negligible.

3) Energy reuse.

The reuse of energy is in fact the use of energy that has been used for our needs and this axis still goes through three ways:

* Reuse of energy that has already been used in the manufacturing process and that has given what is expected of it.

* Reuse products and bring them, as it were, into the circular economy.

* Transforming one form of energy into another, such as sunlight, wind speed, fluid movement, etc.

a) Reuse of the energy already been useful.

In all processes, there is a lot of energy used (Useful) and a lot of energy lost (Recoverable or not recoverable).

This last energy is already studied in the axis of energy efficiency N°2 which is the recovery, on the other hand to recover or more precisely, to reuse the energy already been useful and which gave satisfaction and of which one was happy and satisfied.

If we admit that the useful and already used energy still contains energy that could be reused again, why not try to reuse it, not to forget that it is free?

In fact, this principle already exists in some known industrial processes, I site in what follows some:

* Pasteurization of milk

The pasteurization of milk consists in treating the milk thermally in order to sterilize it partly (in fact, another treatment is envisaged at higher temperatures, approximately of 140°C, called sterilization).

A small recall on the process of pasteurization of the milk, the milk is stored between 4°C and 6°C after being received, by refrigerating circuits with glycolated water (in general), the pasteurization consists in making heat the milk at high temperatures from 70°C to 80°C during a defined time, and as the milk must be stored after this treatment, therefore it is necessary that it is cooled by iced water or glycolated water.

To do this, the milk specialists have designed a special exchanger that includes several phases, not 02 phases for heating and cooling but more in order to reuse the hot pasteurized milk to preheat the milk at the entrance which is cold, thus consuming less heat for the pasteurization phase and also consuming less cold for cooling, since the milk has become less hot.

This technique is known by the dairy plants and allowed to reuse the heat useful for the pasteurization of the product to reduce the needs in heat and cold.

In fact, it should not be confused with heat recovery, since the latter must be done for something lost, but we speak of reuse for something that was useful and is reused another time.

* Compressed Air

In what follows, we talk about the reuse of air, why not? Indeed, the compressed air is mostly used by cylinders, another part is used directly for sandblasting and de-clogging of filters and a third part is used for plastic blowing machines (PET or other).

And of this first and third mode of use that we speak, in fact the compressed air in these two processes, the compressed air is used in the first case, by the jacks to assure movements of translation under the effect of the force exerted by the pressure ($P=F/S$), after which and to assure other cycles, the jacks must return to their initial places and therefore, the air must escape and therefore one must throw it (It gave satisfaction and it was very useful).

And in the third case, at the level of the blowing of the bottles to be made (One can imagine the quantity of the air used equivalent to the volume of all the plastic bottles manufactured by this process), also this compressed air after use, must escape to leave the place to the following cycles.

So we can say that almost all the compressed air produced worldwide with an energy efficiency of 20% is thrown away in nature as such.

It is true that its reuse is in the mind of some energy companies already in the world, but it remains to give value to this reuse, to study and talk about it and to encourage the energy managers to adapt it as an alternative.

The reuse of compressed air is possible, because the compressed air after use does not lose pressure or quality, in general, and for the cylinders, it escapes through an orifice that is linked to the distributor and this through a silencer. The idea, therefore, is to exploit the processes that use compressed air and that are generally cyclic, that is to say, a cylinder in action, another at rest, so it is possible that the air that escapes from the first can be injected into the second and vice versa.

Similarly for bottle blowing, it is necessary to understand the manufacturing cycles and to look for adequate means to reuse compressed air (especially since air is brought to a pressure of 30 bars and more for water filling plants in PET bottles, for example).

b) Product reuse

We hear more and more about the circular economy and that can also be defined as the reuse of products to further extend their presumed lifetimes.

We know that the products and especially those industrialized and not directly consumable such as packaging of all kinds:

- Aluminium

- Wood

- Cardboard

- Glass

- ...etc.

Can be either re-processed into other forms and uses, or recycled. Recycling in fact consists in reusing them even indefinitely.

- Aluminum is recyclable and represents too high a percentage in the waste.

- Wood is also recyclable, especially with the MDF wood technique.

- The cardboards too, which contain the short and long fibers already well dosed.

- Glass is a recyclable product and is highly desired by manufacturers of glass products.

The energy savings are there, even though we still spend energy to recycle them, but because they have already been processed before, it will make the manufacturing and processing chain shorter on the one hand and therefore less energy spent than the one for the transformation of the virgin materials (that's what the manufacturers say anyway) and on the other hand, it preserves the mines, forests and sands whose extraction, transport and pretreatment of the virgin raw materials require energy, water....etc.

c) Energy transformation

The transformation of types and forms of energy into other forms, is well known, for example, the electric motor transforms electrical energy into mechanical energy, the electrical resistance transforms electrical energy into thermal energy... etc.

Now what I mean here is to transform the useful energies for us, for any purpose, into other forms, for example and among the known forms, the sun that shines and that illuminates us our days, this illumination has been well exploited also to produce electric energy (by means of the technology of the semi conductors) that are photovoltaic substances and therefore from the light we will have electric energy.

The same goes for wind speeds, especially at sea or on mountains, where the latter are more or less important and therefore contain kinetic energy that could be transformed into mechanical energy and through alternators into electrical energy.

Another form of natural energy is that of the waterfalls which contains a potential energy (although it is not available by all and where we want), it can also be transformed into mechanical energy necessary for the alternators to be transformed into electric energy.

But I'm going to tell you about other forms and types that are in front of us and that can be reused, since they are already in use, I'm giving some examples to illuminate a new path, in my opinion, it is until now, ignored by many people.

* Exhaust or humid air from dryers or others.

The chimneys of the exhaurs of the dryers or others convey in fact a hot fluid more or less wet but contain two forms of energy: thermal and mechanical, this first form all the world knows, but it is forsaken because it is difficult to recover or there is no interest in its recovery or its recovery is not profitable.

On the other hand, the second form (mechanical) is easy to recover, even more profitable than other types of equivalent recovery, the interest is always there since it can be always transformed into electrical energy which is always useful.

The idea, I do not know if it is clear, is to install wind turbines at the exit of these chimneys and like that we will have an air at relatively constant speed and can be available 24/24 and 7/7, so it is much more profitable than the classic wind turbines whose output is 25% on average.

Examples are to be sought, it is enough to know the philosophy of the paths and tracks of energy efficiency to find others.

That said, the axes and paths are now more or less clear, but it remains to know a little technique and become familiar with the parameters and energy laws.

And for those who are familiar with energy efficiency, I also give you in the following, tracks with quantification of potential energy savings and this to not get lost and save a little on the time I spent in my experience.

Conclusion:

The reuse of already consumed energy is not a phrase with which people are familiar, but we hear more and more about the circular economy and this phrase fits in this context.

However, in order to succeed in energy saving actions, it is necessary to have a good understanding of energy concepts and parameters.

III. Energy parameters

And as I mentioned at the beginning, it is important to know the units of the energy parameters to use.

a) Nuances and Confusion of Energy Units

In this part I will try to provoke the readers by asking questions and I want you to think seriously about the answers, and you will be convinced that many energy parameters are not well understood, others are not well known and maybe others will be heard for the first time.

It is true that knowing a parameter badly or ignoring it may not cause any problems, but sometimes it can lead to abnormal energy consumption and if we go further, it can cause damage and serious dangers.

Before moving on to the energetic parameters which must be very well understood, I will pose other problems, at the level of some physical phenomena that many of us exploit to live well, without being able to explain them.

I start with the phenomenon of drying cloths in the open air, everyone can see that the water initially present in the cloths evaporates little by little at atmospheric pressure, while experiments have been done by us or in front of us at school and which conclude that the evaporation of water at atmospheric pressure can only be done from 100°C. So for the cloths outside, it is absurd that we reach even 60°C. How come, is the physical law missing something? Or are we missing some information?

Another phenomenon is that many people use pressure cookers to cook food more quickly, even though we have the same fire on the stove and the same amount of gas or the same amount of electricity. So what is the energy parameter that comes into play to save us time in the preparation of our food, where does this extra energy come from?

Also, almost everyone knows the balloons that are used in parties and in many events, when they are inflated, they are too light and they can be thrown on, even, the heads of babies and will be without impact, on the other hand by opening their necks, they will be thrown energetically and even adults fear their impact on them, has the question been asked, where does this energy come from that allows the balloon to fly far and hit people? (No energy no displacement).

And many other phenomena that must be explained in order to master the energy consumed, all these phenomena and others, will be assimilated by understanding the energy parameters.

b) Energy parameters and units :

We present the most used units in the energy sector which are part of the International System; SI

Length, width, height: [m], meter.

Surface : [m²], square meter.

Volume : [m3], cubic meter.

Time : [s], second and not the hour, nor the minute.

Power : [W], power.

: [J] (joule) = [W]*[s], not the kJ and not the Energy, work calorie.

Mass : [kg], kilo gram and not the gram.

Force, weight : [N], Newton.

Speed : [m/s] and not km/h
   Static, dynamic, total, relative, absolute pressure: [Pa] (pascal) = [N/m]²

Temperature : [K], Kelvin and not °C

Relative humidity : [%].

Absolute humidity : [g water /kg dry air].

Specific heat (Cp) : [J/kg.K].

Lower Calorific Value, Higher Calorific Value: [J/kg].

Volume : [m3/s], not m3/hr.

Density : [kg/m3],

c) Confusion in the units of energy parameters:

I give in the following, an overview on the probable confusions in the main energy units and we start with the most used.

i) Power

What is the unit of power used in the SI, it is the W (Watt), it is true for those who operate in the field of energy, it can be known, and can be for everyone since energy is the concern of the whole world and I think that the current wars are started because of energy.

On the other hand the powers used are sometimes far from that, we speak of BTU for air conditioners, of Thermie and Calorie for thermal energy, of CV (Horse Power) for engines...etc.

To master energy, it is necessary to know all the possible conversions of these units between them and mainly to arrive at the SI unit (W) which must be obligatorily used in the balances not to be misled (*It is not a law but it is an Expert's Advice*).

Another unit that we are going to see that it can and must be used is the TEP (Tonne Oil Equivalent).

It is necessary to be VERY careful with the sum of the values of the different forms of energy, and it is necessary to say that we cannot always add 1000kWh Electric with 1000kWh Thermal although they are real kWh.

Because, in the majority of the cases, the electric energy is produced starting from the thermal energy and in other cases, starting from the nuclear power and still others starting from the renewable energies (Wind, Hydraulic, Photovoltaic, etc.) and as there are always returns of transformation and as these returns vary much from a type of power station to another what makes that one does not speak about the same primary source of this electric energy, it was thus necessary to speak about another unit for the primary energies which is the TOE.

So, in the calculation of our internal balances of equipment and machines, we add up without any problem, the electric and thermal kWh, while for the balances of factories, countries, globe, it is necessary to pass to the TOE. (Especially that the world currently, speaks a lot of Carbon Balance and that can be calculated only from the TOE)

ii) Pressure

Did you hear about static and dynamic pressure, is there a difference between relative pressure and absolute pressure? If so, what type of pressure is found in the inflated balloon?

Knowing that the word "Static" means immobile and stable (not moving), unlike the word "dynamic", which means movement and mobility.

The word "relative" means that the pressure value is given in relation to the atmospheric pressure, whereas the absolute pressure is given in relation to absolute zero.

So, an inflated balloon, contains a static pressure if it is put on the ground or on the table, but if we open the neck of this balloon, we notice that the balloon starts to move and fly in space, which proves that the pressure has become dynamic.

For pressures (of fluids: gases or liquids), the unit used in the SI is the Pa (**Pascal**) and not the bar (which is the best known unit when talking about pressure), although the Pa is used in air circuits whose pressure is relatively low, (Pa, hPa, kPa), we also find for liquids and especially for water pumping, the mCE (Metre Column of Water).

(1 bar = 10mCE=100000Pa)

Pressure is only a force (N) exerted on a surface (m^2).

$$Pressure = Force / Surface$$

The unit of measurement of pressure, in the international system is the Pa, the most used and best known unit is the bar, so you should know that :

1 bar = 100000 Pa, and 1 Pa= 1N/1m^2

- Relative pressure and absolute pressure,

We also speak of two types of pressure: relative and absolute, it is mandatory to use the latter in the balances and calculations otherwise all our balances will be wrong. The relative pressure is the pressure read on the manometer, it is said to be relative, because we are talking about a value relative to the value of the atmospheric pressure (about 1.013 bar, and actually depends on the geographical coordinates).

While the value of the absolute pressure is given relative to the absolute zero. The pressure is counted from the absolute zero (0 bar absolute) and until the infinite, under the 0 bar absolute, there is absence of pressure and consequently it is the absolute vacuum.

"So don't forget that 1 bar abs = 1 bar rel + atmospheric pressure".

- Static pressure, dynamic pressure, total pressure,

We also talk about other types of pressure, that it is interesting to know them, namely: static and dynamic. The word static means always immobile, like the pressure that is in an inflated balloon put on the ground, since it is inflated so the pressure that reigns there is different from 0 bar and this pressure is called Static.

Contrary to the static word, we have, the dynamic word which means mobility, therefore the movement and the speed, that is to say that the air is in movement, and that we find it in the air which escapes from the balloon after having made a hole inside.

The initial pressure is static, and is therefore transformed into dynamic pressure.

"The total pressure is the sum of the static pressure and the dynamic pressure".

It should be remembered that the presence of a static pressure does not necessarily mean the presence or absence of a dynamic pressure, but any pressure measured is considered as total pressure.

iii) *Temperature*

For the temperature, we work with two types of units, the degree centigrade (°C) and the degree Kelvin (K), the difference is 273.15°C which is the minimum temperature that a substance can reach. So 1K = 1°C+273.15°C.

And if someone uses the °C and another uses the K, they will never have the same result. For example, if two people sell hot water at 50°C, they spend the same energy to heat their water from an initial temperature of 15°C, but the first will sell an energy equivalent to :

$$Ec = \text{Water mass} \times Cp \text{ water} \times 50°C$$

On the other hand, the second will say that he sold an energy equal to :

$$Ek = \text{Mass water} \times Cp \text{ water} \times (50 + 273.15) \text{ K.}$$

So they did not sell the same amount of energy, but it is certainly the same amount of water brought from the same initial temperature to the same final temperature.

One of the two has made a mistake, *it is the first one.*

To conclude, it is necessary to pay attention to the use of these units, in fact, one frequently uses the temperature difference (since one frequently speaks about passage from an initial state to a final state) what will be idem for the two units, in the other case, the error can be catastrophic and thus not to take risk, one will always use the degree Kelvin (K)

Another delicate point is the reference value of the temperature, especially for the room temperature, for example to calculate the losses of hot surfaces in two different rooms that are brought to different room temperatures, the reflex is to use the same reference, or a temperature equal to absolute zero (-273.15°C=0K) or equal to 0°C.

However, the losses are a function of the actual value of the room temperature, so we must be careful with this.

iv) Humidity

The humidity also enters the energy balances because the air is in general wet and thus contains water in the form of vapor, the relative humidity measured in (%), is a concept not used directly in the energy balances on the other hand it is it which is measured to define another concept of humidity which is that absolute defined in (g water/kg dry air)

In fact, we can not measure the absolute humidity directly, but through the digram of the humid air and measuring in general the relative humidity and the dry temperature (or both dry and humid temperatures). This diagram is essential to know and master.

In fact, humidity is a very important parameter, since on the one hand, water vapor contains a certain amount of energy (depending on its temperature and pressure), which must be taken into account in the balances and on the other hand.

Similarly, high absolute humidity means that there is a lot of water vapor in the air.

IV. **Optimization of the Energy Performance.**

I quickly present a list of possible actions in different consumption areas in order to help in the reflection on the rational use of energy.

a) Energy saving in electrical installations:

- Reduce the electrical voltage.

- Compensate reactive energy with capacitor banks.

- Minimize the rate of voltage drop in the cables.

- Substitute LED lights for lighting.

- Think SMART lighting.

- Replace standard motors with more efficient IE2 and IF3 motors.

- Install variable speed drives on motors whenever the load is variable.

- Properly size the motors and adapt them to the load of the machines.

- Turn off equipment whenever it is not needed.

- Automate and control the switching on and off of lights.

- Seriously think about replacing the traditional servo-controls and regulations (valves, flaps or bypass) by an Electronic Variable Speed Drive regulation.

- Promote new technologies that save energy and do not focus only on the price, the purchase price is sometimes negligible in front of the expenses on energy.

b) Energy saving in compressed air systems:

- Reduce pressure losses in compressed air networks.

- Reduce the operating pressure to the minimum necessary.

- Install variable speed drives on air compressor motors.

- Reduce the pressure setting range.

- Think of making more than one pressure network whenever the needs of the equipment are very different.

- Close the network.

- Automate the operation of air compressors.

- Schedule leak detection and repair campaigns.

c) Energy saving in refrigeration systems:

- Reduce pressure losses in water networks

- Install variable speed drives on air compressor motors and chiller pumps

- Minimize HP

- Increase the BP as much as possible

- Protect air-cooled condensers

- Refresh the cooling air of the condensers

- Favour water cooled condensers when hot water is required (see Mixed condenser)

- Think of making more than one network with different temperatures, whenever the needs of the equipment are very different.

d) Energy Saving in Steam Plants:

- Use steam pressure as needed.

- Install variable speed drives on burner motors and feedwater pumps

- Reduce the temperature of the flue gases as much as possible.

- Control and minimize the amount of oxygen required for combustion.

- Automatically control the deconcentration purge.

- Treat make-up water with at least softened water and at best osmosed water.

- Periodically check traps and valves for leaks.

- Favour food tarpaulins of the pressure type to the atmospheric one.

- Insulate all steam valves as well as pipes.

- Recover as much steam condensate as possible.

- Reduce steam pressure as much as possible.

- Check the tubes and heat exchangers for scaling and descale them whenever necessary.

Figure 2: Heat loss rate as a function of scale thickness

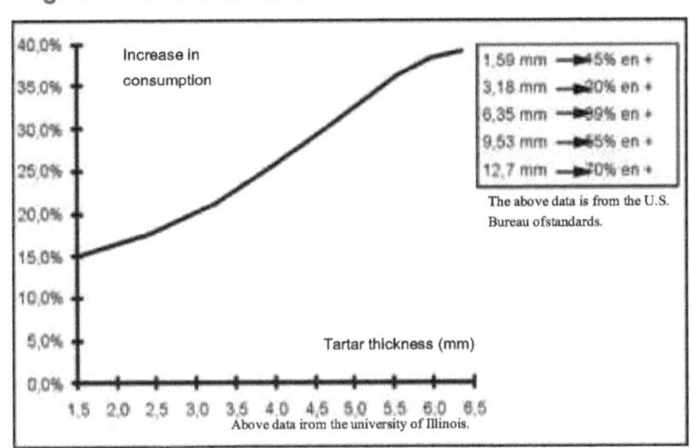

e) Energy saving in air conditioning installations:

- Consider installing individual air conditioning units for CEOs, GMs, ministers and whenever the position is specific.

- Think of units with INVERTER.

- Centralize and control the fan convectors to be able to control them.

- Always think about Free Cooling.

- Sometimes favor the extraction and supply of new air than the air conditioning of technical rooms.

- Consider using both forms of energy available at the condenser and evaporator of air conditioning units.

- Consider recovering the energy available at the air compressors and condensers of refrigeration production units for space heating.

- Pay attention to the type of insulation of air-conditioned building envelopes whenever the internal load is so high.

V.   Potential of Energy Gains.

a) Main industrial energy consumption items

- Manufacturing process.

- Compressed Air.

- Industrial refrigeration (Iced water, Glycol water, Negative cold, Cold rooms, cooling towers...etc.)

- Heating (Steam, Hot Water, Thermal Fluid...etc.).

- Air conditioning (Blowing, Extraction, Filtration...etc.)

- Pumping, Ventilation & Extraction.

- Lighting.

- Various

b) Share of each item in the energy bill

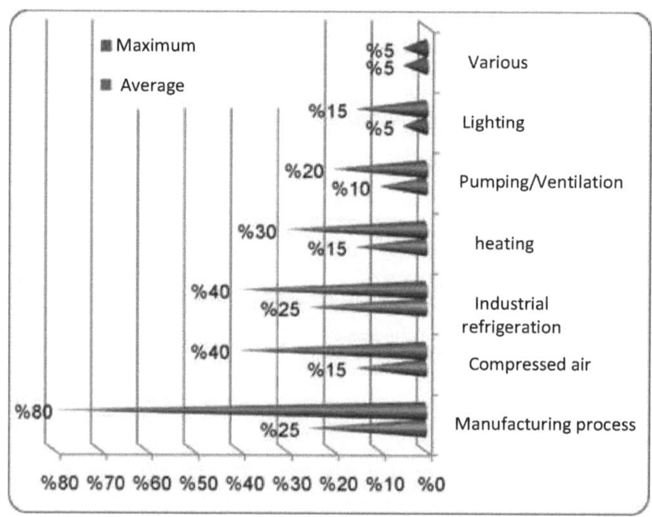

Figure 3: Consumption rate by energy item (Expert Ali BEN AOUIENE)

c) Energy savings by consumption item

i) Compressed Air

- The share of compressed air in the energy bill is on average 15%. (Min: 5%, max: 35%).

- The cost of energy represents 75% of the compressor's lifetime expenses, while the cost of purchasing represents only 12% to 13%.

- An increase of 10°C in the air intake temperature leads to an increase in power consumption of 3.8%.

Figure 4: Air compressor

- 1 bar less in the operating pressure is equivalent to a 10% saving (7 bar compressor).

- There is no need to compress the air to 7 bars and then expand it to 2 or 4 bars, it is more economical to make more than one network at different pressures.

- A looped compressed air network is equivalent to dividing pressure losses by 2.

ii) Heating and steam

- 0.8mm of scale on the tubes of a boiler = 8.2% of fuel loss. (3.2mm of scale is 25% loss).

- 1% more in combustion efficiency=1% less fuel consumption.

Figure 5: Steam boiler

- The re-evaporation of the condensate at 8 bars, in the atmospheric food tank, represents 12.5% of this condensate.

- One ton of condensate at 8 bar discharged contains 20% of the energy of one ton of steam.

- Insulating the valves leads to significant energy savings, especially in cold periods (more than 5% gain).

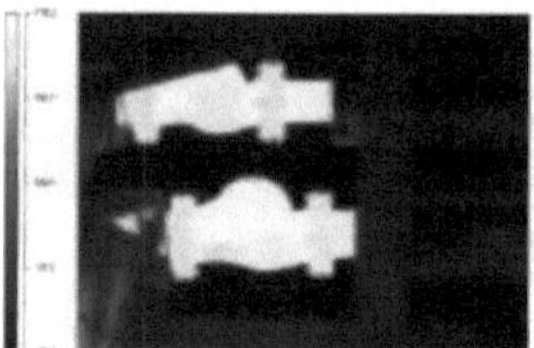

Figure 6: Real case: The temperature at the valves is about 172°C due to the lack of insulation

- Minimize the blowdown rate to just what is needed by targeting a steam boiler water conductivity of 3000microsiemens/cm. (gain more than 5% if automatic salinity control)

- Minimize excess air in the flue gas to 3% for natural gas for example.

iii) Air conditioning

- Reducing the air conditioning temperature by 1°C is equivalent to saving 6-8%.

- Reduce the pressure drop in the pipes as much as possible, the power of the motors is proportional to the square of the pressure drop.

- Reduce the mixing rate (Air Change Rate) to just what is necessary, the power of the motors is proportional to the flow rate conveyed.

- Choose the right type of filtration and systematically control the clogging of the latter.

- Increase the chilled water temperature, where possible, to reduce energy-intensive dehumidification (the condensation of water vapor from the air).

iv) Pumping & Ventilation

- The efficiency of a fan or pump is optimal if its operating point is in the middle of its characteristic curve.

- The average efficiency of centrifugal pumps is only 55% (medium efficiency) to 80% (high efficiency)

- Always have the highest possible suction pressure and the lowest possible discharge pressure.

- Reduce the pressure drop in the pipes as much as possible, the energy loss is proportional to the square of the pressure drop.

v) Electric Motors

- The motors must be operated at 75% of their loads for maximum efficiency.

- A high efficiency motor has an efficiency of 90-96% while a standard efficiency motor has an efficiency of 84-92%.

- During the life of the engine, energy-related expenses represent 95% of its overall expenses.

- If the flow rate is variable, a variable speed drive is much more economical than a damper or a control valve. (Savings of 87% versus 12% for the same action and the same output flow).

- Reactive energy compensation by capacitor banks (5% of the electric bill)

Figure 7: capacitor banks

Figure 8: Power factor controller

vi) Lighting

- The lighting represents in the industry, on average a share of 5% (from 2% to 15%).

- It is recommended to replace standard lamps with new technologies.

- LED; >50% saving / Standard lamps; life span: 50000h vs. 10000hLStd.

- Neon type T5; 30% saving / neon type T8 and 16000h for the T5 against 10000h for the T8.

- The least efficient lamps are incandescent lamps. (15lm/W) (Lifetime: 1OOOh)

- The most efficient lamps are LED lamps **(>200lm/W)** (5OOOOh).

- Fluorescent lamps have an efficiency of 1OOlm/W (lifetime: 1OOOOh).

d) Usual specific consumption

In general, the overall (Specific Consumption) ratio in the industry is not very significant contrary to the ratios of the equipment or consumption item.

- Reference ratios of some utilities:

- COPc of refrigeration compressors: > 3

- 700 th to produce one ton of steam at 8 bars.

- 130Wh/m3 of compressed air at 7 bars.

- >100lm/W for lighting lamps.

- Central heating system efficiency: > 65%.

- ***Fuel combustion efficiency***

    - Natural gas: 90%- 95%.

    - Heavy Fuel Oil 2   : 85% -90

- ***Ratios: Textile Industry***

- Washing   : 0,8kWh/Piece

- Clothing (jeans):   0.5 kWh/piece

- Dye   : 0,6kWh/kg

- Spinning   : 4kWh/kg

- Weaving, treatment and finishing: (Fabric) 1.5kWh/ml & 5 th/ml.

- ***Ratios: Paper Industry***

- Paper   : Electric 350-700kWh/ton

- Paper   :Thermal 1500-3000kWh/tonne

- ***Ratios: Aluminum Industry***

- Processing & lacquering: 450-550kWh/t

- Machining and lacquering: 20kWh/m^2.

- ***Ratios: Miscellaneous Industries***

- Electrical cables   : 500kWh/t

- Brick   : 40kWh/t and 400th/t

- Glace   :700kWh/m3et300th/m3

- **Energy savings potential by consumption item:**

- On average, the energy saving potential in industrial plants can reach **30% and more.**

- The minimum to be targeted per company is **10 to 15%.**

- In the compressed air station, the gains can reach more than **50%.**

   * Reduction of the operating pressure of air compressors (1 bar = 10% gain)

   * Reduction of compressor idling (15%-50% gain)

   * Reduction or even elimination of compressed air leaks (average gain of 30%)

- In the cold station, the possible gains are **30% and more**

   * Installation of variable speed drives on variable loads (gains of 20%-80%)

   * Optimization of the chilled water production system. (gains of 10%-25%)

- In the steam area, the potential savings are on average **15%.**

   * Combustion improvement **(5-10% gain)**
   * Automatic control of water salinity in the boiler **(5% gain on blowdown)**

   * Insulation of pipes and valves **(gains 4-5%)**

   * Descaling of heat exchangers **(0.8mm=8.5% gain)**

- In the Lighting item, the expected gains are more than **50%.**

- In the air conditioning sector, the expected gains are more than **30%.**

- In electrical installations, the possible savings are **5%.**

   * Optimization of the Electricity Billing System, (**monetary gains 8% to 10%)**

   * Optimization of the Natural Gas Billing System (**monetary gains 5% to 8%)**

   * Harmonic attenuation by filters (**gains of 2% to 3%)**

*   Reduction of voltage drop in cables **(gains of 1% to 3%)**

Conclusion

In this book, it has been specified the great axes and paths that can be traveled to ensure a rational use of energy, axes that are summarized in the optimization of the performance of the energy, the recovery of the energy already lost and the reuse of the energy that has already been useful.

However, I think that the first step to follow is to master the units of energy parameters, then comes the understanding and training on the maximum of physical and technical phenomena (All that moves necessarily contains an energy), namely: Electrical, Mechanical, Thermal, Chemical...etc.

Then, it is necessary to be able to estimate, simulate or quantify the potential of energy saving, not to forget that the optimization, the recovery or the reuse of several forms of energy is possible, remains to define the how to define and to quantify well the saving and the investment.

Finally, take action, preserve our environment, reduce and/or recover losses and balance the functioning of the earth.

Specific books on energy efficiency by station and by utilities will be published: (Compressed Air, Cold, Steam & Heat, Pumping & Ventilation, Air Conditioning, Electrical Installation...etc.)

yes
I want morebooks!

Buy your books fast and straightforward online - at one of world's fastest growing online book stores! Environmentally sound due to Print-on-Demand technologies.

Buy your books online at
www.morebooks.shop

Kaufen Sie Ihre Bücher schnell und unkompliziert online – auf einer der am schnellsten wachsenden Buchhandelsplattformen weltweit! Dank Print-On-Demand umwelt- und ressourcenschonend produzi ert.

Bücher schneller online kaufen
www.morebooks.shop

info@omniscriptum.com
www.omniscriptum.com